School Publishing

PATHFINDER EDITION

By Rebecca L. Johnson

CONTENTS

Leapin' Lizar

By Rebecca L. Johnson

What's that leaping into the air?
Is it a furry mouse? A slimy frog?
No, it's a lizard with scaly skin.

ds

Lizards are amazing animals. They can crawl, run, and climb. Some lizards can swim. And some are pretty good jumpers. Lizards come in many shapes, colors, and sizes. The world's smallest lizard can fit on a dime. The largest is longer than the tallest basketball player.

But just what is a lizard, exactly? Scientists **classify** living things into groups based on certain characteristics, or **traits.** Lizards are **vertebrate** animals. That means they have a backbone—just like you. Lizards are also **reptiles.** Snakes, turtles, crocodiles, and alligators belong to the reptile group, too.

Scaly Skin

Different kinds of animals have different body coverings. Most mammals have fur. Birds have feathers. Frogs and most other amphibians have skin that's slippery and moist.

Dry, hard **scales** cover the bodies of reptiles. Scales are tough like your fingernails. Scales cover a lizard's body, from the end of its nose to the tip of its tail. Some scales are thick and bumpy. Others are smooth. Scales may overlap as shingles on a roof do.

Other Lizard Traits

What other traits do lizards have? Lizards have tails. Most have four legs and feet with five toes. Many lizards also have claws.

Unlike snakes, lizards have ears you can see. They usually look like little holes on both sides of a lizard's head. When a lizard opens its mouth, you can see lots of sharp teeth.

Most lizards have eyelids. They blink and close their eyes when they sleep. Geckos are lizards that don't have eyelids. Their eyes are covered by large, clear scales instead.

Quick Change Artist. This chameleon is a type of lizard that can change colors to blend into its surroundings.

Self-Cleaning. Geckos like this one use their tongues to clean the scales on their eyes.

Egg Layers

Most lizards lay eggs on land like other reptiles. The eggs have tough, leathery shells. Lizards lay eggs in sand, under dead leaves, or in some other warm, safe spot. In a few types of lizards, the eggs hatch inside the mother's body. Then she gives birth to live young.

Lizards don't stick around to see their eggs hatch or babies grow. When baby lizards are born, they are completely on their own.

A World of Lizards

Looking for lizards? It isn't hard to find them. Worldwide, there are about 5,000 different kinds. Lizards live on every continent except Antarctica. Many islands are home to lizards, too.

Desert Dwellers

Gila monsters live in deserts. They are found in the southwestern United States and northern Mexico. These lizards are heavy and slow moving. They also have a poisonous bite.

Gila monsters eat the eggs of birds and other lizards. They use their strong front legs to dig eggs out of the ground.

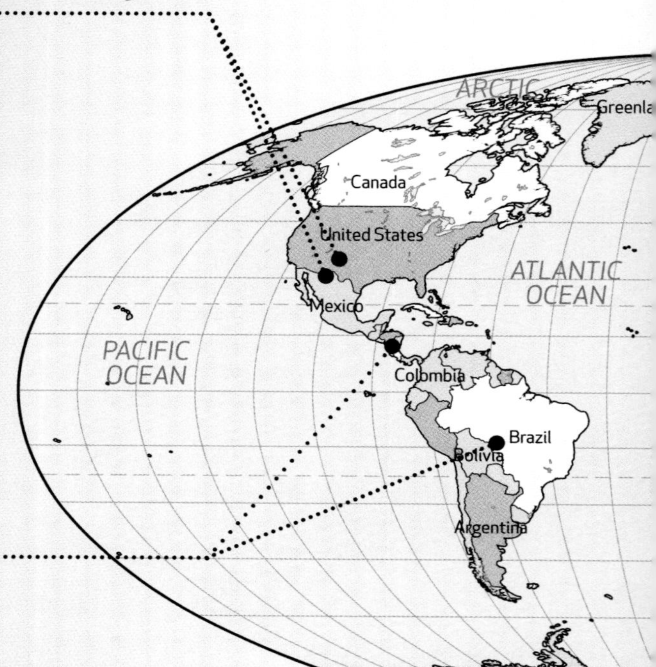

Up in the Trees

Walk through a rain forest in Central or South America and you'll see green iguanas up in the trees. These climbing lizards have long, skinny toes tipped with sharp claws. They have pointy scales along their backs.

Iguanas are tough. If they fall out of a tree, they don't get hurt. They are good swimmers, too.

Talented Toes

Geckos are small lizards. They are found in many warm parts of the world—too many to point out on the map! Geckos can grip any surface with the special pads on their toes. Some kinds of geckos live in people's homes. People usually like having them around because geckos eat insects.

Red Heads

South Indian rock agamas live in southern India. They eat mostly insects. They are fast runners with strong, long legs.

These lizards' brown color makes them hard to see when they sit on rocks. However, male rock agamas change color during part of the year. Their heads and tails turn bright red.

What a Tongue!

Northern blue-tongued skinks live in forests and grasslands in northern Australia. They have plump, heavy bodies. Unlike many lizards, blue-tongued skinks have weak, stubby legs.

When danger threatens, these lizards don't run away. They just stick out their bright blue tongue. That's usually enough to scare an enemy away.

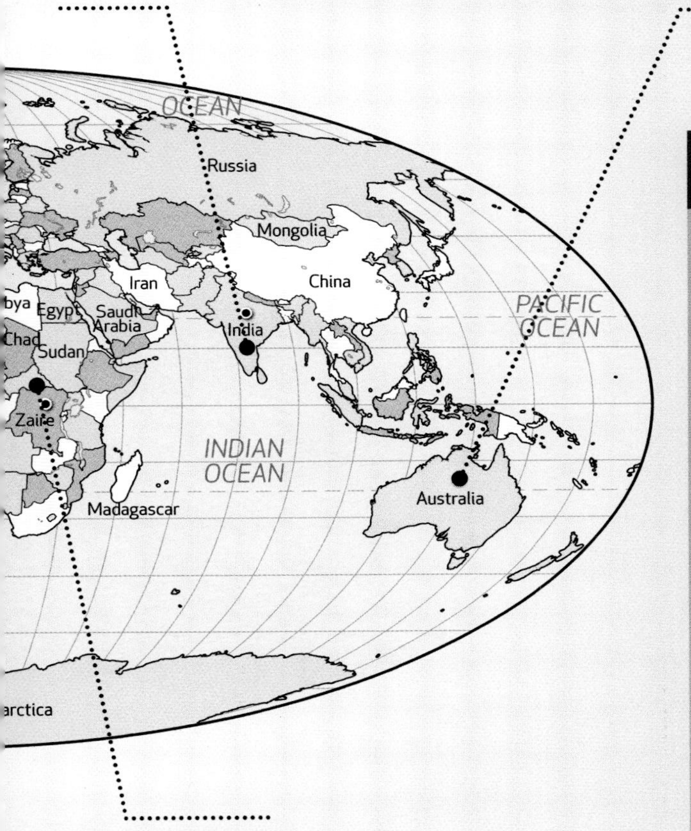

Wordwise

adaptation: a behavior or body part that helps an animal survive

classify: to group living things with similar traits

reptile: a group of air-breathing vertebrates that includes lizards and other animals whose bodies are covered in scales

scales: tough structures that cover the bodies of lizards and other reptiles

trait: a feature or behavior of a living thing

vertebrate: an animal with a backbone

Special Sight

Many kinds of chameleons live in Africa, including the island of Madagascar. A chameleon's eyes stick out from its head. Each eye can look in a different direction. This **adaptation** lets a chameleon look in two places at once!

Chameleons move slowly. So how do they catch the insects they eat? Chameleons have incredibly long tongues with sticky tips. With lightning speed, a chameleon can shoot out its tongue to snag a meal.

Great Escapes

Lizards have all sorts of adaptations for staying alive in a dangerous world. Many adaptations help lizards escape predators or avoid being eaten. Take a look at the lizards and their adaptations below.

Don't Touch

A thorny devil's body is covered with sharp, spiky scales. The spikes usually keep predators from taking a bite. If the spikes aren't enough, the thorny devil can shoot blood from its eyes. It can hit a predator a meter (3 feet) away. That's enough to scare almost anything off.

A Trick Tail

Many lizards have a tail that will break off if a predator yanks on it. The enemy is left with a wriggling tail while the lizard escapes. A new tail grows back in about a year.

Fancy Feet

The basilisk lizard has skin between its long toes. It has webbed feet, like a frog. This adaptation lets the lizard run on water. It can zip across a stream or pond to make a quick getaway.

The Rock Trick

The armadillo lizard doesn't run from danger. Its body is covered with thick, sharp scales. When trouble comes, the lizard grabs its tail in its mouth. It curls up into a spiky ball that looks like a sharp-edged rock. Who would want to eat that?

Twin Traits

Lizards have many special traits. Things like dry, scaly skin make them different from other animals. But lizards and other animals are alike in some ways, too. Two different kinds of animals can have similar adaptations that help them survive.

Chameleons and monkeys both have tails that can wrap around a branch and hold on. These grabbing tails work like an extra hand. They leave all four feet free to do other things.

Sometimes it helps to be colorful. The anole lizard does this by pushing out a flap of red skin under its chin. The frigate bird does it by filling a red sac on its neck with air.

Looking bigger and scarier can make predators turn and run. Both the frilled lizard and the cobra have a similar trick. They can make their heads look bigger than they really are.

No wings? No problem! A flying dragon lizard has flaps of skin between its front and back legs. So does a flying squirrel. Both these animals spread their skin flaps to glide through the air.

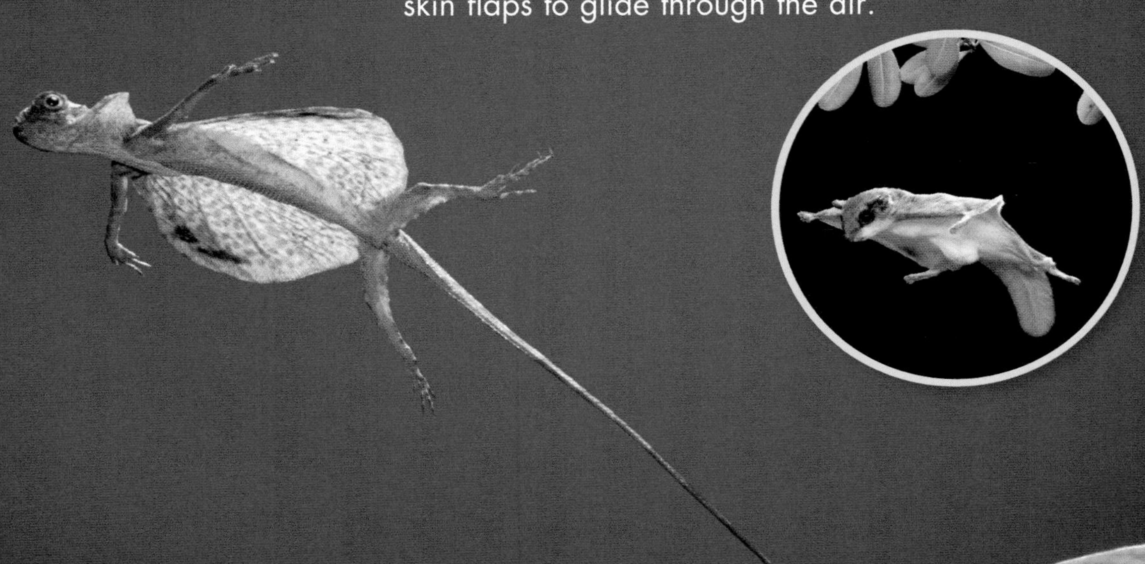

Leapin' Lizards

Show what you've learned about lizards by answering the questions below.

1. Why do scientists classify a lizard as a vertebrate?
2. Name three traits that lizards have.
3. Choose two of the lizards shown on pages 6 and 7. How are they alike? How are they different?
4. What is an adaptation?
5. Give an example of an adaptation that a lizard and a different kind of animal share.

INDEX

Acknowledgments
Grateful acknowledgment is given to the authors, artists, photographers, museums, publishers, and agents for permission to reprint copyrighted material. Every effort has been made to secure the appropriate permission. If any omissions have been made or if corrections are required, please contact the Publisher.

Photographic Credits
Cover ZSSD/Minden Pictures; 2-3 ZSSD/Minden Pictures; 4-5 Stephen Dalton/Minden Pictures, Kathy Tschoerner/National Geographic Image Collection; 5 Modoki Masuda/Minden Pictures; 6 Jim Merli/Visuals Unlimited, Claus Meyer/Minden Pictures/National Geographic Image Collection, Daniel Heuclin/NHPA/Photoshot; 7 Ganesh H Shankar/Alamy, Gary Bell/OceanwideImages.com, Thomas Marent/Minden Pictures; 8-9 Theo Allofs/Corbis; 9 R. Andrew Odum/Peter Arnold Inc./Alamy Images, Joe McDonald/Visuals Unlimited, Daniel Heuclin/Biosphoto/Peter Arnold Inc.; 10 Ian Nichols/National Geographic Image Collection, Roy Toft/National Geographic Image Collection, Art Wolfe/www.artwolfe.com, James H. Robinson/Photo Researchers Inc.; 11 Belinda Wright/National Geographic Image Collection, Joe McDonald/Visuals Unlimited, Oxford Scientific (OSF)/Satoshi Kuribayashi/Photolibrary, Satoshi Kuribayashi/Minden Pictures; 12 John Foxx Images/Imagestate.

Illustrator Credits
6-7 Precision Graphics.

Program Authors
Randy Bell, Ph.D., Associate Professor of Science Education, University of Virginia, Charlottesville, Virginia; Malcolm B. Butler, Ph.D., Associate Professor of Science Education, University of South Florida, St. Petersburg, Florida; Kathy Cabe Trundle, Ph.D., Associate Professor of Early Childhood Science Education, The Ohio State University, Columbus, Ohio; Judith Sweeney Lederman, Ph.D., Director of Teacher Education and Associate Professor of Science Education, Department of Mathematics and Science Education, Illinois Institute of Technology, Chicago, Illinois; David W. Moore, Ph.D., Professor of Education, College of Teacher Education and Leadership, Arizona State University, Tempe, Arizona

National Geographic School Publishing
Hampton-Brown
www.NGSP.com

Printed in the USA.
National Graphic Solutions, Appleton, WI

ISBN-13: 978-0-7362-7714-3

11 12 13 14 15 16 17 18 19
10 9 8 7 6 5 4 3

Product #4E90219

ISBN 978-0-7362-7714-3

NATIONAL GEOGRAPHIC

School Publishing

Physical Science

PATHFINDER EDITION

NATIONAL GEOGRAPHIC

Explore

ON YOUR OWN

Sound All Around

By Rebecca L. Johnson

It's Time to Explore on Your Own!

Good readers use multiple strategies as they read on their own. Use the four key reading comprehension strategies below:

PREVIEW AND PREDICT

- Look over the text.
- Form ideas about how the text is organized and what it says.
- Confirm ideas about how the text is organized and what it says.

2 **MONITOR AND FIX UP**

- Think about whether the text is making sense and how it relates to what you know.
- Identify comprehension problems and clear up the problems.

MAKE INFERENCES

- Use what you know to figure out what is not said or shown directly.

SUM UP

- Pull together the text's big ideas.

Remember that you can choose different strategies at different times to help you understand what you are reading.